Robots

Written by Jonathan Emmett

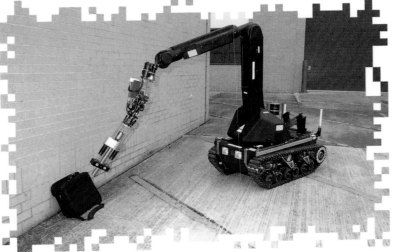

Contents

What is a robot? 2
Tiny robots and big robots 4
What robots look like 6
What robots do 8
Inside a robot 14

Collins

What is a robot?

A robot is a machine that does a job for you.

Robots can do many jobs.

Tiny robots and big robots

This tiny robot is as small as a coin.

This robot is bigger than a man.

This robot is very strong.

What robots look like

Some robots look like people.

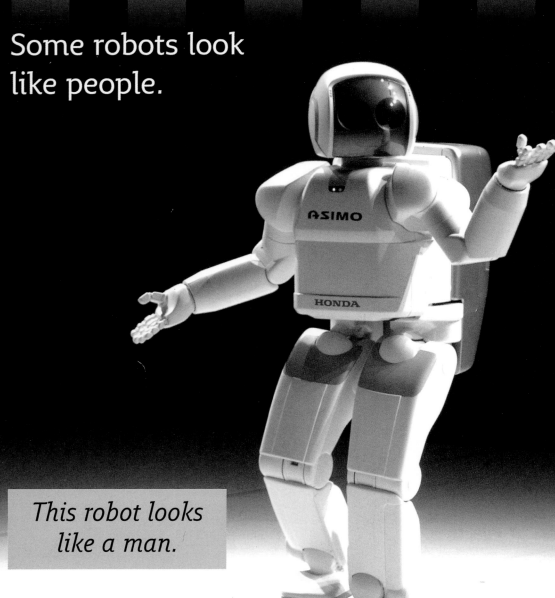

This robot looks like a man.

Some robots look like animals.

This robot looks like a dog.

This robot looks like a shark.

What robots do

Some robots go to places where people can't go.

Some robots go to the bottom of the sea ...

This robot dives deep.

... or to another planet, like Mars.

These robots are on Mars.

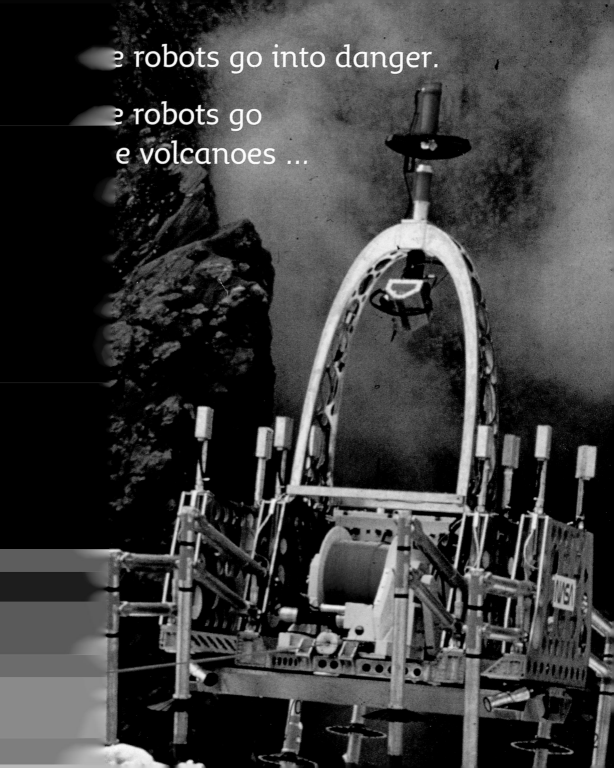

... and some robots look for bombs.

Some robots work hard ...

These robots are building a car.

... and some have fun.

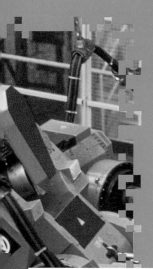

This robot wants to win.

Inside a robot

Camera: the robot uses this to see

Motor: this makes the robot move

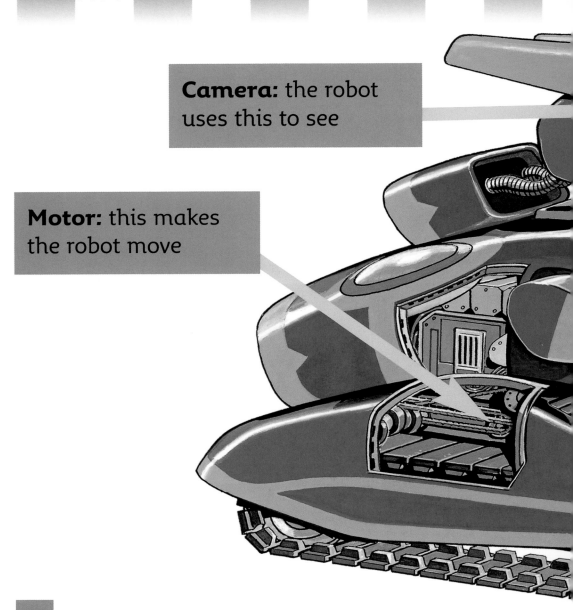

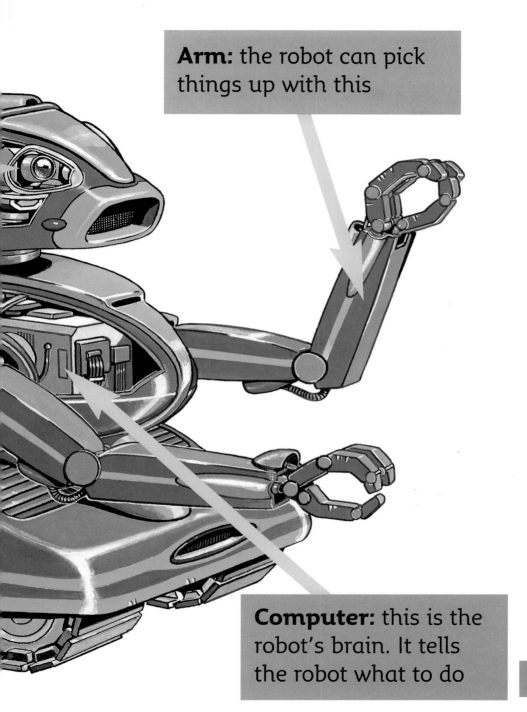

Arm: the robot can pick things up with this

Computer: this is the robot's brain. It tells the robot what to do

Ideas for reading

Written by Alison Tyldesley, MA PGCE
Education, Childhood and Inclusion Lecturer

Learning objectives: using terms fiction and non-fiction and understanding their different features; predicting text, reading on, leaving a gap and re-reading; reading words with initial consonant clusters; taking turns to speak and listen.

Curriculum links: Science: Ourselves (parts of the body)

High frequency words: is, that, big, this, than, where, can't, or, another, some, have

Interest words: machine, people, animals, planet, danger, volcanoes, bombs

Word count: 106

Resources: small whiteboards and pens

Getting started

- Read the title and discuss what the book is about. Ask the children what they already know about robots, and discuss how this book is about robots in real life.
- Ask what the proper word is for books that tell us facts. Use the words 'fiction' and 'non-fiction'.
- Walk through the book with the children, looking at the pictures, and discuss the different types of robots they see.
- Ask the children what they should do if they get stuck on a word. Model reading a sentence and leaving a gap for the problem word, then returning to see what word 'fits'.

Reading and responding

- Ask the children to read the book up to p13, aloud and independently. Observe and prompt where appropriate, and praise children who predict unfamiliar words by leaving a gap and then finding the word that fits.
- Ask children, in pairs, to look through the text and find examples of words with an initial consonant cluster, e.g. *sm*all p4, *pl*aces p8, *pl*anet p9. Help the children to blend both initial sounds.
- Look at pp14–15 together and discuss. How is this illustration different from one in a fiction book? What kind of illustration is it? (It is a diagram with